Eratosthenes and His Sieve

by A. I. Freeman
illustrated by Wendy Born Hollander

Printed in China

ISBN 978-1-328-77185-8

3 4 5 6 7 8 9 10 0940 25 24 23 22 21 20 19 18

4500700961 A B C D E F G

Eratosthenes's Achievements

Eratosthenes was a man of many interests. He was born in 276 BC in Cyrene, in North Africa. Eratosthenes began his learning in Cyrene. Then he went to Athens, Greece, to study under great teachers.

Ptolemy III, the king of Egypt, noticed him. The king asked Eratosthenes to go to Alexandria to tutor his son. Before long, Eratosthenes became the head of the Library of Alexandria. He spent the rest of his life in the city.

Eratosthenes wrote a book about geometry.

He made a list of all the known stars.

He measured the distance of the Sun and Moon from Earth.

He measured Earth's tilt.

He worked on a calendar that included leap years.

He wrote long poems.

Eratosthenes did all this and more, but he is best remembered for two things.

Eratosthenes is best remembered for measuring Earth and showing that it is round. His measurement was very close to the actual size of Earth.

He is also remembered for his sieve of prime numbers. A sieve is a strainer. It has tiny holes or a screen. Big things stay on top of the screen and smaller things fall through. It separates by size. Eratosthenes's sieve didn't separate by size, but it did separate different kinds of numbers. Let's see how it works.

Read·Think·Write For which two things is Eratosthenes best remembered?

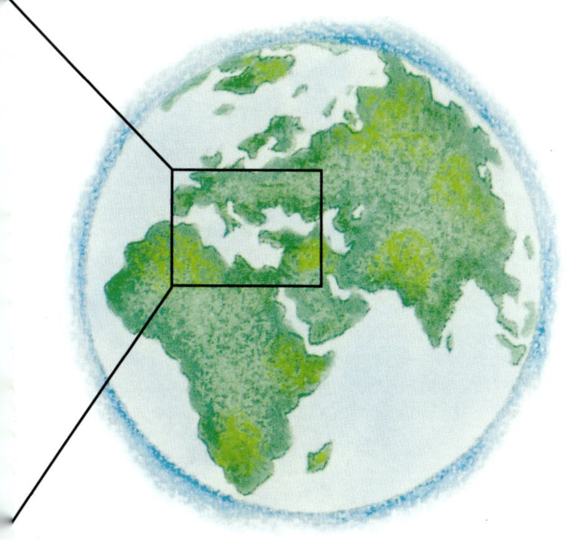

Eratosthenes thought that Earth was round. He even measured it!

Prime and Composite Numbers

A prime number is a whole number that has only itself and 1 as factors. Here are some examples:

The number 2 is a prime number. It has only itself and 1 as factors.

The number 3 is a prime number. It has only itself and 1 as factors.

The number 11 is a prime number. It has only itself and 1 as factors.

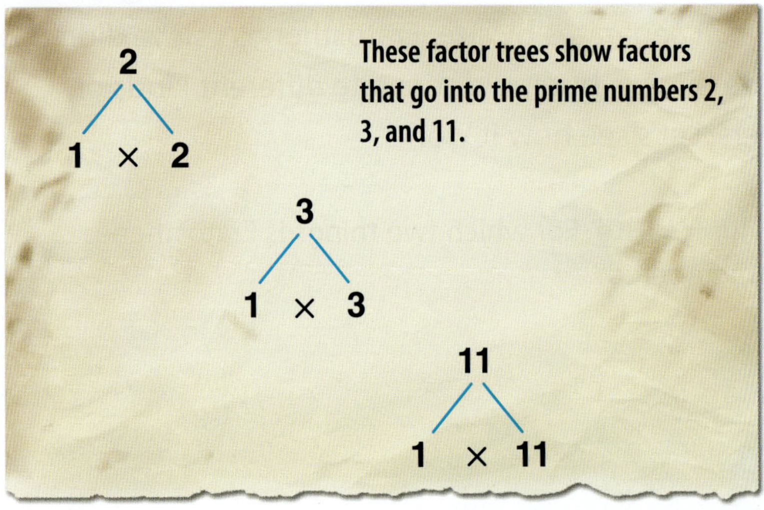

These factor trees show factors that go into the prime numbers 2, 3, and 11.

Read·Think·Write Why is 5 a prime number? Is 7 a prime number? Why?

A composite number is a whole number that has more than two factors. The number 4 is a composite number, because it has the factors 1, 2, and 4.

Except for 2, all even numbers are composite numbers. Do you know why? The number 2 is a factor for all even numbers, such as 4, 6, 8, and so on. So, all even numbers other than 2 have at least three factors.

Odd numbers can be composite numbers, too. For example, the factors of 9 are 1, 3, and 9. The factors of 15 are 1, 3, 5, and 15.

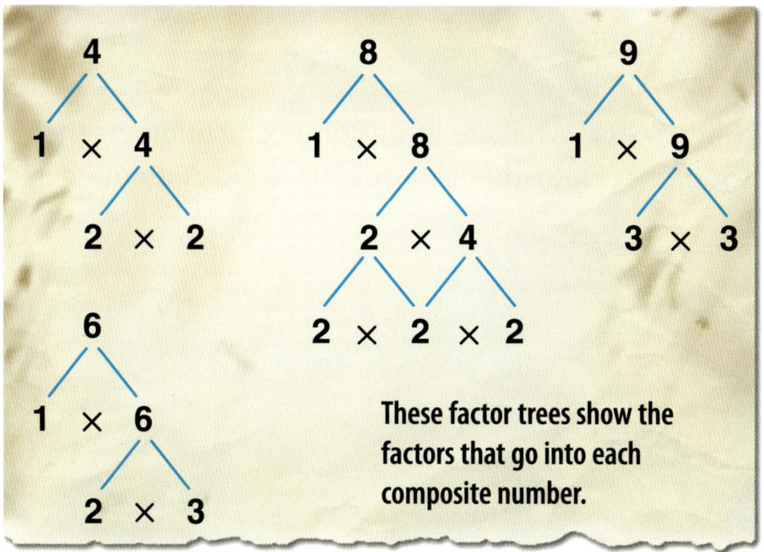

These factor trees show the factors that go into each composite number.

Read·Think·Write Is the number 10 a composite number? Why?

The Sieve of Eratosthenes

Eratosthenes wanted an easy way to figure out prime numbers. First, he wrote out a list of numbers. He crossed out 1. The number 1 is neither a prime nor a composite number. It only has one factor.

Then he circled 2 because it's a prime number. He crossed out all the other even numbers, because 2 is a factor of all even numbers.

Next, he circled the prime number 3. Then he crossed out all the odd numbers that 3 is a factor of. The even numbers that 3 is a factor of had already been crossed out.

Read·Think·Write What are three composite numbers that 3 is a factor of that Eratosthenes would have crossed out next?

~~1~~	2	3	~~4~~	5	~~6~~	7	~~8~~	~~9~~	~~10~~
11	~~12~~	13	~~14~~	~~15~~	~~16~~	17	~~18~~	19	~~20~~
~~21~~	~~22~~	23	~~24~~	~~25~~	~~26~~	~~27~~	~~28~~	29	~~30~~
31	~~32~~	~~33~~	~~34~~	~~35~~	~~36~~	37	~~38~~	~~39~~	~~40~~
41	~~42~~	43	~~44~~	~~45~~	~~46~~	47	~~48~~	~~49~~	~~50~~

When Eratosthenes had crossed out all the numbers that 3 is a factor of, he moved on to 4. He didn't have to cross out any numbers. Do you know why? The number 4 is an even number. The number 4 can be divided by 2. So, any number that can be divided by 4 is an even number that had already been crossed out.

Eratosthenes continued on to number 5. He crossed out numbers that 5 is a factor of. If you know how to count by fives, this is an easy one. He crossed out any number that 5 is a factor of that hadn't already been crossed out. That means he crossed out 25, 35, and 45. Eratosthenes continued in this way to make his sieve.

People are still working on finding new prime numbers. Today we can use computers to help sieve big, long prime numbers.

Read·Think·Write After 5, what number did Eratosthenes circle next? Then what numbers did he cross out?

Responding Math Concepts

1. What happens if you multiply any prime number by the prime number 2?
2. What happens if you multiply any prime number by the prime number 5?
3. What happens if you multiply any prime number by the composite number 4?
4. Which of these is a composite number, 71 or 81?
5. What is the next prime number over 100?

Activity

Draw Conclusions Work with a partner to discuss these questions. Use the table on page 7 to help you.

- How can you describe prime numbers?
- How do you know that the number 18 is not a prime number?
- How can you tell that the number 37 is a prime number?
- What can you say that describes all composite numbers?

PHOTOGRAPHY CREDITS: **Cover** © Dennis Hallinan/Alamy; **2** © Detlev van Ravenswaay/Picture Press/Getty Images; **4** © NASA; **5** © NASA; **6** © Oleg Nikishin/Getty Images

Printed in China

ISBN 978-1-328-77200-8

3 4 5 6 7 8 9 10 0940 25 24 23 22 21 20 19 18

4500700961 A B C D E F G

What would you do if you had an extra $20 million? If you were Dennis Tito, you would buy a ticket to go into space!

Read·Think·Write How would you write twenty million as a whole number?

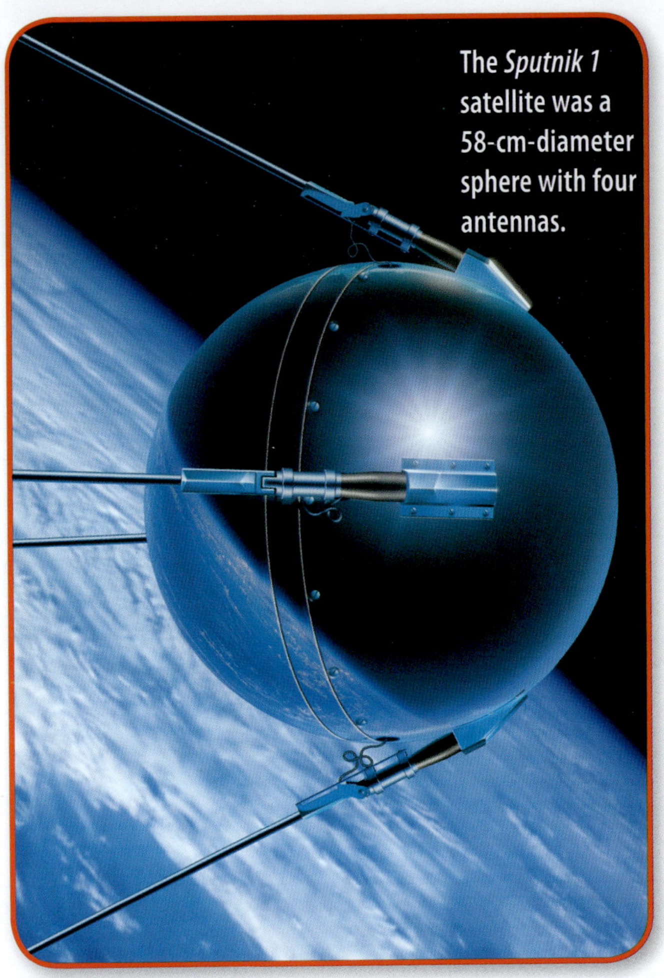

The *Sputnik 1* satellite was a 58-cm-diameter sphere with four antennas.

Dennis Tito was born in Queens, New York, in 1940. His father was a printer. His mother was a seamstress. Tito was a teenager when *Sputnik* was launched in 1957. The Russian *Sputnik* was the very first satellite ever sent into space. The memories stayed with Tito for years. That experience was the reason he got a degree in astronautics in college. It was also the reason he went to work for the National Aeronautics and Space Administration (NASA).

At NASA, Tito worked on designing trajectories for space missions. Trajectories are the paths, or orbits, that the missions take in space. Tito stayed at NASA for five years. Then he went to work on Wall Street in the stock market.

Tito used what he knew about engineering and math to develop new approaches to investing in the stock market. After a few years, he opened his own business in California. By age 40, Tito had made his first million dollars. A few years later, that same business was worth nearly $200 million.

Read·Think·Write How many digits are in the number 200 million?

Although he wasn't in the space business anymore, Tito still was very interested in it. On a trip to Moscow in 1991, he convinced the Russian government to let him ride along on a supply ship. The ship was going to *Mir*, a Russian space station 248 miles above Earth. Tito began training with the cosmonauts, or Russian astronauts. But the trip was called off. Russia had decided to stop using the 15-year-old space station. Soon after that, *Mir* crashed to Earth.

Read·Think·Write In the number 248, which digit is in the hundreds place? the tens place? the ones place?

Tito was later offered a ride on another Russian ship. Its mission was to take supplies to the new International Space Station (ISS). The ISS was only about two years old. Tito said yes, but NASA said no. What did NASA have to do with it? The new ISS had been built with the cooperation and money of 16 countries. However, the United States and Russia had the most say in its planning. They also put in the most money. Because of this, NASA felt they had a right to object to Tito's plan.

The International Space Station

Dennis Tito poses with the International Space Station crew on April 30, 2001. Tito is on the right in the middle row.

NASA was very worried about safety. In 1986, teacher Christa McAuliffe was onboard the space shuttle *Challenger*. Shortly after takeoff, *Challenger* blew up. McAuliffe and the whole crew were killed. Because of what had happened to the space shuttle, Tito had to work hard to win NASA over. Eventually he did. NASA said he was a one-time exception.

Tito had to agree that if he broke anything he would pay for it. He was not allowed to go anywhere alone on the ISS. He also had to sign a paper that said he would not hold Russia or the United States responsible for what might happen to him if another disaster occurred.

On Saturday, April 28, 2001, 60-year-old Dennis Tito's dream came true. For eight days, he got to be an astronaut. Six of those days were spent orbiting Earth—almost 16 times per day, or about 94 orbits altogether.

After his trip, Tito went back to his business in California. He likes opera, sailing, and fast cars. He encourages people to do things they've always wanted to do, and he believes that everyone should be able to travel into space.

Do you have a dream? What would it take for you to make it come true? If you're anything like Dennis Tito, you will do whatever it takes!

Read·Think·Write Dennis Tito made about 94 orbits of Earth while in space. How do you write that number in words?

Responding

Math Concepts

1. How do you write ten million dollars, using numbers?
2. Which number is in the hundreds place in 5,384?
3. How would you write the number three thousand six hundred eighty-two?
4. Categorize and Classify Which number is in the tens place in the number in question 3?

Activity

Distances in space are very big. They are measured in AUs instead of miles. AU stands for astronomical unit. An AU is the average distance from the Sun's center to Earth's center. An AU is equal to 92,955,807 miles.

Make a place-value chart that includes millions, thousands, and ones. Write the number of miles in AUs on your place-value chart.

Use markers or crayons to outline the millions part of your chart in blue, the thousands part of your chart in green, and the ones part of your chart in red. Share your chart with a partner.

Multiplying a Good Deed

by Caroline Ross
illustrated by Doreen Gay-Kassell

Printed in China

ISBN 978-1-328-77201-5

3 4 5 6 7 8 9 10 0940 25 24 23 22 21 20 19 18

4500700961 A B C D E F G

"Those sandwiches look yummy," said Ramon's grandfather as he walked into the kitchen. "But why do you look so serious?" He and his grandson had lunch together every Saturday. Today was Ramon's turn to cook.

"Look at all this food, Grandpa!" Ramon said. "We always have something to eat. You said that some people in our community go to bed hungry because they don't have enough money for food. That makes me so sad. I wish I could do something about it."

"Maybe you could organize a food drive at school," said Grandpa. "Think how much food you'd collect if everyone donated just one can!"

"That's an awesome idea, Grandpa! I'll talk to my teacher about it."

Ramon could hardly wait to get to school the next day. His teacher, Ms. Hill, listened closely as Ramon laid out his plan.

"I like your idea," she said, "but it will take a lot of hard work. You'll need to find people to help you and figure out how to let everyone know about your plan. You'll also need to find a place to store all those cans and a way to get them to the food pantry. Are you willing to meet all those challenges?"

"No problem!" answered Ramon. "I thought we could ask the students in the school to each contribute one can of food. There are two classes in each grade, and the fourth grade has about thirty kids per class. That's a lot of food from just our grade alone!"

Read•Think•Write About how many students are in the fourth grade at Ramon's school?

Ramon's classmates all wanted to help. Ms. Hill found an empty classroom in which to store the cans. Grandpa arranged for a truck to pick up the cans and deliver them to the food pantry.

The class decided they would collect cans for two weeks. If each student at school brought just one can, they would have more than 300 cans of food.

Ramon spoke to the students in each grade. In a second-grade class, one girl asked, "Could we each bring two cans of food?"

"What a terrific idea!" said Ramon.

Read•Think•Write How many cans would the 48 students in second grade donate if everyone brought two cans?

Ramon thought about what the second-graders had decided to do. What if he asked students in every grade to bring two cans instead of one? Better yet, what if he asked them to bring three or four cans? How many cans of food would his class collect then?

Ramon made a chart. First, he wrote the number of students in each grade. In his head, he rounded the number of students to the nearest ten. Then he multiplied by 2 and 3 to get estimates. Finally, he multiplied the actual number of students by 2 and 3. He wrote the answers in the last two columns.

Grade	Number of Students	2 cans	3 cans
Kindergarten	48	96	144
1	54	108	162
2	48	96	144
3	53	106	159
4	47	94	141
5	59		

Read•Think•Write How many cans of food would the fifth-graders bring if they each brought two cans? Three cans?

Ramon shared his findings with his classmates and asked what they thought. He even estimated how many cans there would be if each student brought four or five cans. "We know there are 309 students in the school. If I round 309 to 300 and then multiply by 4, I get 1,200 cans of food! If I multiply by 5, I get 1,500!" he exclaimed.

In the end, Ramon's class asked students to bring as many cans as they wanted. Each grade kept a record. It turned out that the fourth grade brought in the most cans!

Read•Think•Write How many cans of food would there be if each student in Ramon's school brought four cans?

The students packed the cans in boxes, with 24 cans in each box. There were 20 boxes of soups and 9 boxes of stews. There were also several boxes of canned fruits and vegetables.

When all the boxes were packed, workers from the food pantry picked up the cans. They were grateful for such a large donation.

"I'm very proud of you," Grandpa told Ramon. "You saw a need and you did something about it."

"Yes," smiled Ramon. "Great things can happen when you multiply a good deed!"

Read•Think•Write How many cans of stew were collected?

Responding

Math Concepts

1. Summarize There are 54 first-grade students in Ramon's school. How would you round the number 54 to the nearest ten?
2. There are 59 fifth-graders. Estimate the number of cans the students donated if they each brought four cans of food.
3. What is the actual product of the problem above?
4. Many students brought large cans of fruit. There were seven servings of fruit in each can. Estimate the number of servings in 128 cans.
5. How many actual servings of fruit were in the cans?

Activity

Write a problem in which a two-digit number is multiplied by a one-digit number. Trade your problem with a partner. Use base-ten blocks to show how to solve each other's problem. Remember that you may need to regroup the ones.

by G. Ramirez
illustrated by Michelle Dorenkamp

Printed in China

ISBN 978-1-328-77202-2

3 4 5 6 7 8 9 10 0940 25 24 23 22 21 20 19 18

4500700961 A B C D E F G

In two weeks, the Midway Lions would be on the program *School Quiz* on channel 48. Chase, Asa, Tricia, and Anthony were members of the team. For the past few weeks, they had practiced for the quiz.

"I think we should practice division with remainders next," said Asa.

"What are remainders?" asked Anthony.

Tricia wrote 35 divided by 4 on the board. "I'll explain by showing you a problem," she said. "Let's say I want to divide 35, the dividend, by 4, the divisor. To start, I ask myself what I need to multiply by 4 to get an answer close to 35."

"Four times 9 is 36," answered Anthony.

"Since the dividend is 35, 9 groups of 4 are too many. But it's close, so I'll try 8. Eight is my quotient," continued Tricia. "I multiply the quotient by the divisor and get 32. When I subtract 32 from 35, I get 3. Three is the remainder."

"You're right, Asa," said Anthony. "I'm rusty on division with remainders. We'd better practice."

"I know a game we can play to practice," said Chase. "Well, it's more like a math puzzle."

"Good," said Anthony. "I love to solve puzzles."

"Okay!" said Chase. "Your job is to find the dividend. The dividend is the same for all these math statements. If the divisor is 2, the remainder is 1. If the divisor is 4, the remainder is 1. If the divisor is 5, the remainder is 1. If the divisor is 3, there is no remainder."

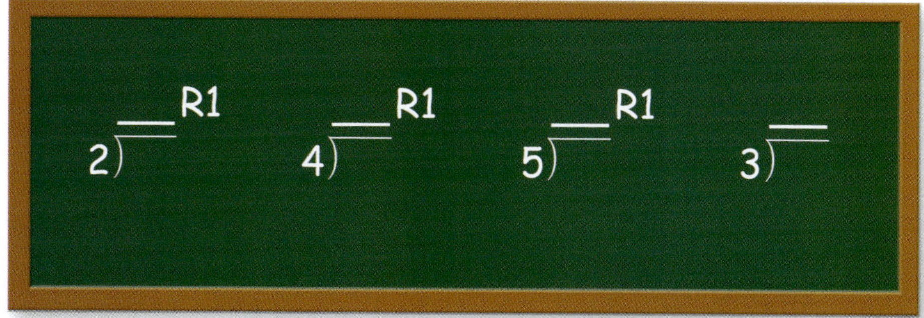

Read·Think·Write What is a remainder?

"And remember," said Chase, "the remainder is always less than the divisor."

"Let's start with what we know," said Tricia.

"The number is a multiple of 3. I know this because there is no remainder if the divisor is 3," said Anthony. "Let's try products that can be divided by 3."

"Good idea!" said Asa. "Three times 2 is 6. Six divided by 2 won't work because there is no remainder."

"You're right," said Anthony. "Maybe 9 will work."

"Nine divided by 5 won't work, either, because there is a remainder of 4. The remainder should be 1," said Tricia. "I think the answer is 21. It can't be 12, because 12 is divisible by 2 and 4. There isn't a remainder. It can't be 15, because 15 is divisible by 5 with no remainder. Eighteen is also divisible by 2."

"Tricia solved it!" said Chase. "Can someone make up another puzzle?"

"I've got one!" said Anthony. "How many tacos did a cook make? When 5 kids ate an equal number of the tacos, 4 tacos were left over. If 7 kids ate an equal number of tacos, there would be 3 left over. If 6 kids or 8 kids ate an equal number of tacos, there would be none left over."

"Okay," said Asa. "The dividend can be divided by both 6 and 8 with no remainder. Twelve can be divided by 6, but not by 8 with no remainder. Also, 18 can be divided by 6, but not by 8 with no remainder. Oh! I think it's 24."

"If 24 is divided by 5, there is a remainder of 4!" exclaimed Tricia. "And if 24 is divided by 7, there is a remainder of 3! Anthony, is the answer 24 tacos?"

Anthony grinned. "I'll tell you on the way to the taco shop. All this taco talk has made me hungry."

Read•Think•Write Was Asa correct? Show your work.

The day of the quiz show finally arrived. The Midway Lions felt nervous, but they did fine.

The Fairview Eagles had also been practicing. For each question Midway got right, Fairview got one right. Finally, the buzzer sounded and the game was over. Midway tied Fairview with 100 points. It was time for the tiebreaker question.

The announcer read the question. "I'm thinking of a number. If it is divided by 6, there is a remainder of 2. If it is divided by 4, the remainder is 2. If it is divided by 3, the remainder is 2. If it is divided by 2, there is no remainder."

Read•Think•Write What is the answer to the tiebreaking question?

The Midway team members smiled. Then Anthony hit the buzzer. He looked at each of his team members, and then they all answered at the same time, "Fourteen!"

"That's right!" the announcer cried.

The Midway Lions had won!

After the show was over, some members of the Fairview team asked the Lions how they knew the answer. Chase got out pencil and paper and showed them how to divide with remainders.

$$\begin{array}{r} 2 \text{ R2} \\ 6\overline{)14} \\ -12 \\ \hline 2 \end{array} \qquad \begin{array}{r} 3 \text{ R2} \\ 4\overline{)14} \\ -12 \\ \hline 2 \end{array} \qquad \begin{array}{r} 4 \text{ R2} \\ 3\overline{)14} \\ -12 \\ \hline 2 \end{array} \qquad \begin{array}{r} 7 \\ 2\overline{)14} \end{array}$$

Responding

Vocabulary

1. Which number is the divisor?

$$\begin{array}{r} 5 \text{ R}1 \\ 3\overline{)16} \\ -\underline{15} \\ 1 \end{array}$$

2. What number is the quotient? $4\overline{)44}$

3. One of the prizes for the winning team is a box of 38 CDs. The Midway Lions team members split the CDs equally. How many CDs did each member get? How many were left over?

Activity

Visualize Use a 10 × 10 grid and colored markers to solve:

$2\overline{)43}$ $5\overline{)47}$ $7\overline{)29}$ $3\overline{)94}$

Hint: Color rows to represent tens. Color squares to represent ones. Divide the tens by the divisor. Divide the ones by the divisor. Remember, a remainder is less than the divisor.

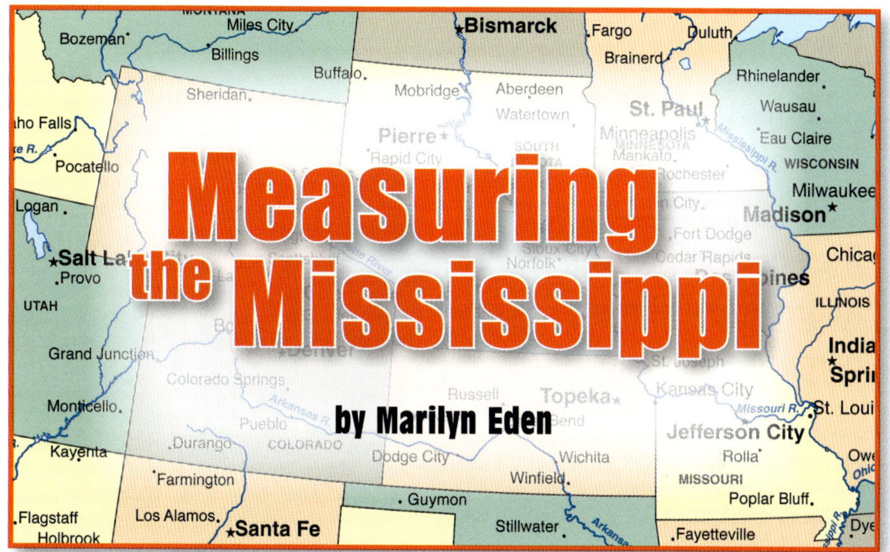

PHOTOGRAPHY CREDITS: **Cover** © f11photo/Shutterstock; **2** Courtesy of the Library of Congress, LC-USZ62-5513; **5** © Getty Images; **6** © franco lucato/Shutterstock

Printed in China

ISBN 978-1-328-77203-9

3 4 5 6 7 8 9 10 0940 25 24 23 22 21 20 19 18

4500700961 A B C D E F G

All Aboard!

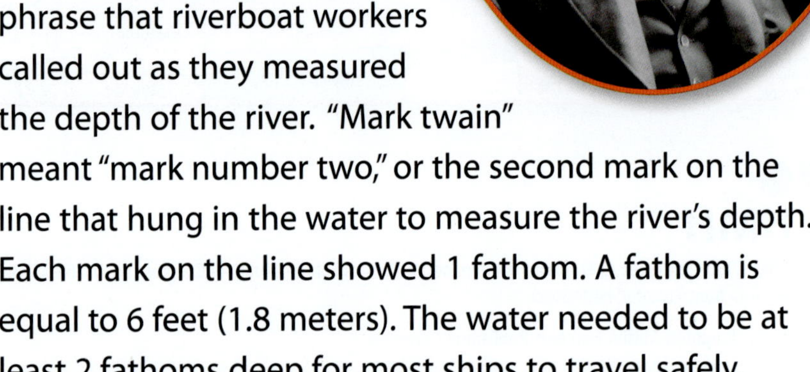

Famous writer Samuel Clemens grew up in Hannibal, Missouri, along the Mississippi River. You might know him by his pen name: Mark Twain.

He took this name from a phrase that riverboat workers called out as they measured the depth of the river. "Mark twain" meant "mark number two," or the second mark on the line that hung in the water to measure the river's depth. Each mark on the line showed 1 fathom. A fathom is equal to 6 feet (1.8 meters). The water needed to be at least 2 fathoms deep for most ships to travel safely.

Today you can experience the same excitement and sense of history that young Sam Clemens knew. Take a paddle-wheel boat ride on the Mississippi River!

Read·Think·Write Which is longer, 1 foot or 1 meter?

About Your Cruise

On your Mississippi River cruise, you will leave from the Twin Cities in Minnesota. Here the river is bordered by bluffs that rise 200 to 300 feet (60 to 90 meters) on each side.

Heading south downstream, your paddle wheeler will go through a series of 29 locks and dams and travel 850 miles (1,368 kilometers) before it reaches St. Louis, Missouri. Here you will see the beautiful Gateway Arch. The arch is a monument to early settlers, who got their supplies in St. Louis before starting the difficult trip west in covered wagons.

You will continue south to Memphis, Tennessee, where you can hear blues music on the streets where it was born. Then you'll stop at Vicksburg, Mississippi, to see a Civil War battlefield. Your trip will end in New Orleans, Louisiana, which is also where the river ends. Here you can eat delicious food while you watch the mighty Mississippi flow into the Gulf of Mexico.

Read·Think·Write Which is longer, 1 mile or 1 kilometer?

About the River

The Mississippi River begins at Lake Itasca, in Minnesota's North Woods. Here the river is small enough to wade across. It's hard to imagine that this small stream grows to become the longest river in the United States.

The Mississippi River is 2,350 miles (3,782 kilometers) in length, stretching from Minnesota all the way to Louisiana. At its mouth in New Orleans, the river flattens out and spreads into a fan-shaped delta. Then it empties into the Gulf of Mexico.

Read·Think·Write
On this map, how many inches is it between St. Louis and New Orleans? How many centimeters?

St. Louis, Missouri

The Ojibway people called the river *Messipi*, which meant "Big River." It was also called *Mee-zee-see-bee*, or the "Father of Waters." Native Americans traveled the river on canoes and hunted along its banks. European settlers later made the Mississippi an important shipping route. Today barges still move goods from Midwest factories and farms down to the Gulf of Mexico. There the goods are loaded onto ships that travel all over the world.

The Mississippi River is also a great place to have fun on the water. Speedboats, fishing boats, jet skis, canoes, and kayaks all share the river with the barges and paddle wheelers. Along the shore, people enjoy bird watching, hiking, camping, and biking. Several scenic highways along the river connect historic river towns.

About the Boat

You will ride on the *River Queen*. She was built like a steamboat that traveled the river in the 1800s.

The paddle wheel has 64 boards that churn through the water. Two diesel engines provide the power that turns the wheel. On the original boat, this power would have come from a steam engine.

That's not the only improvement that's been made to the newer model. A swimming pool has been added. The pool is marked in customary measurements (yards) for passengers from the United States and in metric measurements (meters) for everyone else!

Read·Think·Write Which system of measurement do you use?

When you're not busy eating, swimming in the pool, or enjoying the scenery, you might want to try one of these other popular activities:

- Fly a kite from the deck. You'll be supplied with a kite and several yards (meters) of string.
- Listen to stories about life long ago on the Mississippi River. The boat's Mark Twain storyteller looks and sounds like the real deal!
- Make a carnival mask. You'll be given enough beads and sequins to cover every inch (centimeter) of your mask with sparkle!
- Enjoy a calliope concert. This old-fashioned musical instrument is powered by steam, just like the old steamboats used to be.

Whether you're interested in beautiful views, American history, water travel, or all of these things, you'll find something to love on a Mississippi River paddle wheeler cruise. Book your tickets today!

Read·Think·Write Which would be shorter: something that is 1 yard long or something that is 1 centimeter long?

Responding

Vocabulary

1. Which system of measurement is most common in the United States: customary or metric? Which system is most common in the rest of the world?
2. Predict/Infer If you were going to travel from one state to another state, would you measure the distance in centimeters, meters, or kilometers?

Activity

These animals are all found in the Mississippi River.

Freshwater mussel

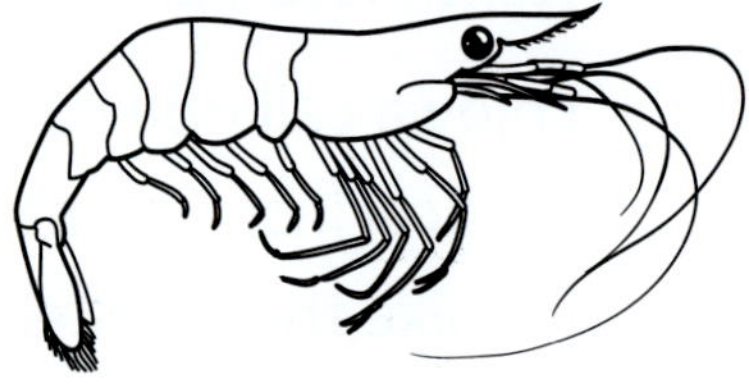

Glass shrimp

Silver lamprey

Use a ruler that shows both inches and centimeters to measure each one.

Skateboarding Takes Shape

by Loretta West

PHOTOGRAPHY CREDITS: **Cover** © Joe McBride/Getty Images; **1** © goldenange,2008/Shutterstock; **2** © Lambert/Getty Images; **10** © Jay Reilly/UpperCut Images/Getty Images; **11** © ZUMA Press, Inc./Alamy

Printed in China

ISBN 978-1-328-77204-6

3 4 5 6 7 8 9 10 0940 25 24 23 22 21 20 19 18

4500700961 A B C D E F G

Surfers to Skateboarders

Imagine racing down a steep hill while balancing on nothing more than a board with four wheels. Sound like fun? The very first skateboarders thought so!

The sport of skateboarding began in Southern California in the 1950s.

The first skateboarders were called "sidewalk surfers." They were surfers with a problem. They needed something fun to do when the ocean was calm and they couldn't surf.

The goal of these "sidewalk surfers" was to create the same thrills and chills on land as they found in the ocean. So they'd start at the top of a hill and ride down on their homemade skateboards. They mostly hoped to stay on their boards and make it to the bottom without crashing into anything—like a car, a tree, or another skateboarder!

By the 1970s skateboarding was very popular. Because there were so many skateboarders in the United States, it made sense that they would want a special place to perform their stunts. They wanted a place to skate where they wouldn't have to worry about broken sidewalks and busy streets.

Skateparks were—and are—the perfect solution! Skateparks provide skateboarders with a safe yet exciting place to skateboard. Most skateparks have a variety of ramps. They have small, gently angled ramps for beginners. And they have steeper and more challenging ramps for experienced skaters.

Skateparks use angles like the ones below to create ramps. They are both acute angles because they are less than 90°. A smaller angle makes a perfect ramp for beginning skateboarders.

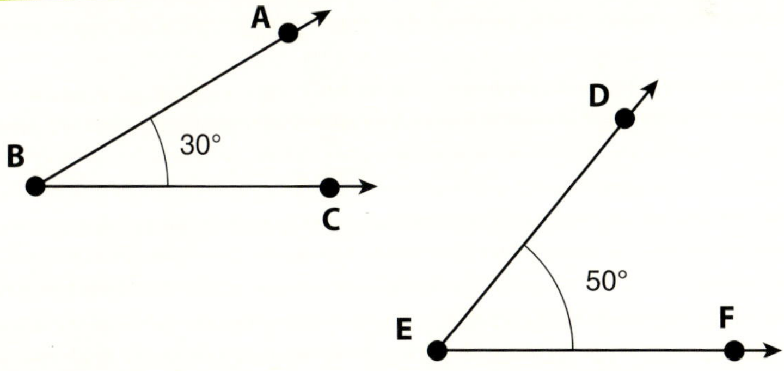

Read·Think·Write Which is the smaller angle? Which is the steeper angle?

Special Places to Skateboard

You probably don't even realize it, but most of the ramps, rails, and platforms found in a skatepark are based on geometric shapes. Look at the picture. A triangle is a three-sided shape. A rectangle is a four-sided shape. Ramps are made by combining triangles and rectangles. Circles are also used in skateparks. Half-pipes and quarter-pipes are formed by using parts of a circle.

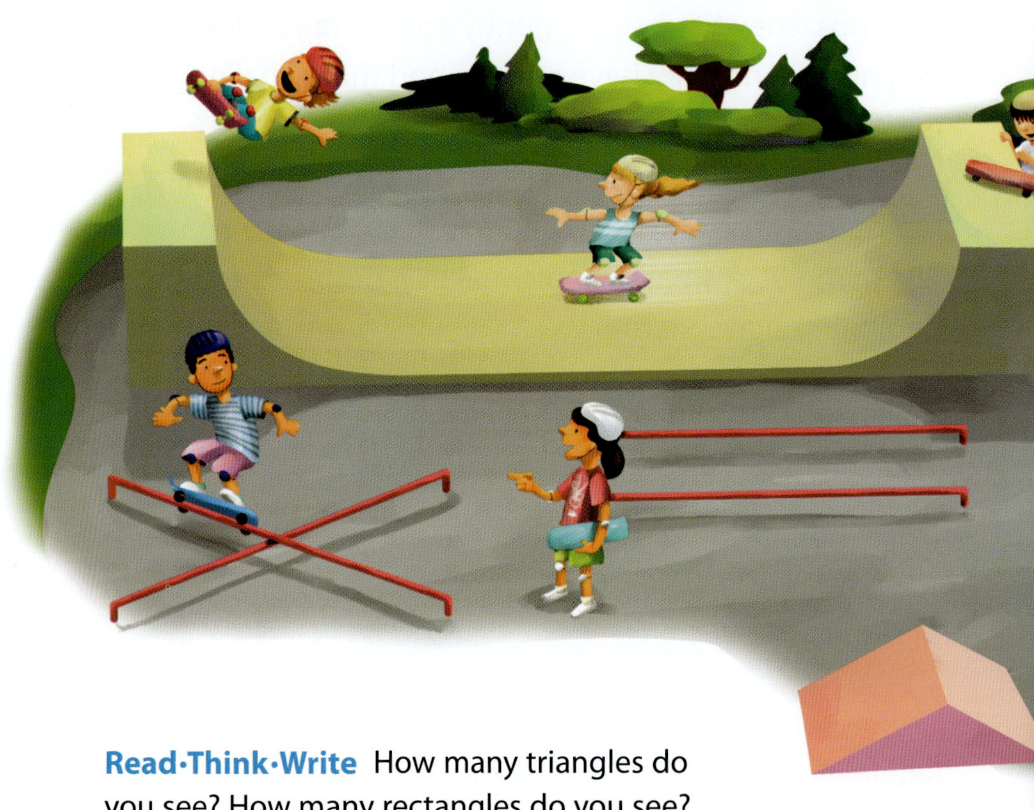

Read•Think•Write How many triangles do you see? How many rectangles do you see?

Parallel lines can also be seen in any skatepark. Parallel lines are always the same distance apart, and they never cross. They are different from intersecting lines, which do cross each other. Parallel lines are also different from perpendicular lines, which meet to form a right angle. By using these different lines, a skatepark can offer skateboarders a number of places to skateboard, such as rails, benches, and grind boxes.

For Beginners

Let's take a closer look at the kinds of ramps a beginning skateboarder might use.

This is a launcher ramp. It's about 3 feet high, 4 feet wide, and 7 feet long. It's not very steep, but it's steep enough to get you started.

Read·Think·Write What geometric shapes make up a launcher ramp?

This ramp is called a quarter-pipe. It's rounded like part of a circle.

The shape below is called a grind box. It can be used to join a launcher ramp and a quarter pipe. Or it can be used alone to practice leaps and jumps.

Read·Think·Write What geometric shapes make up a grind box?

More Advanced

Like all athletes, skateboarders have to practice, practice, and keep practicing some more to be able to leap and flip over obstacles. Some skateboarders are so good that they resemble acrobats as they fly through the air, making complete circles while still attached to their boards. As you watch them perform you might even think, "How did they do that?"

As much as their stunts have to do with balance, they also have to do with the kinds of ramps skateboarders use. One of the ramps a more experienced skateboarder might use is a half-pipe.

Several ramps can be put together to make a funbox. A funbox element is made of different polygons. A polygon is a flat shape with three or more straight sides. Triangles and rectangles are polygons. Take a close look at this funbox. What is the shape of the top of the platform in the center? How do you know?

Read·Think·Write Is every rectangle a square? Is every square a rectangle?

Twist, Turn, and Sail Through the Air!

It's pretty amazing what a person can do with a skateboard and a few ramps, isn't it? Since the first "sidewalk surfers," skateboarders have pushed the limit on using ramps, rails, and half-pipes.

Alan "Ollie" Gelfand invented the ollie in 1978. The ollie has become one of the most common stunts in skateboarding.

Lyn-Z Adams Hawkins started skateboarding when she was very young. She wanted to be like her big brother. At the age of thirteen, Lyn-Z won silver and bronze medals at the 2003 *X Games*. She went on to win another silver medal at the 2005 *X Games*.

Responding

Vocabulary

Look at the skateboarders and their skateboards to answer the following questions.

1. Which skateboarder is skating on a ramp that has a rectangular top?
2. Which skateboarder is performing a stunt on a ramp that looks like a half-circle?
3. Which skateboarder's arms are parallel with the skateboard?

A

B

C

Activity

Visualize Draw your own skatepark or funbox. Include at least one triangle, one rectangle, and a set of parallel lines. Label each of these in your drawing.

A Melody in Fractions

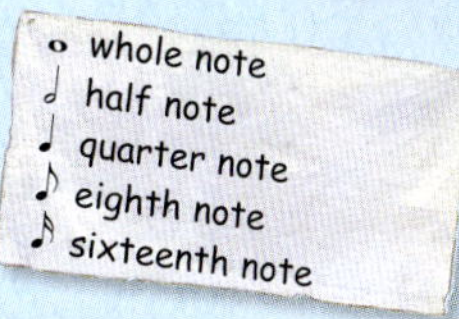

by Ellen Philips
illustrated by Nancy Lane

Printed in China

ISBN 978-1-328-77205-3

3 4 5 6 7 8 9 10 0940 25 24 23 22 21 20 19 18

4500700961 A B C D E F G

Sachiko's heart was racing. Today was her first piano lesson, and she could hardly wait to get started. Sachiko had dreamed of playing an instrument for a long time. She had promised her parents that she would work hard and listen to her teacher.

As she rang the doorbell, melodies popped into Sachiko's head. "Some pieces are slow," she thought as she hummed a favorite tune. "Each note seems to last forever. Others go so fast I bet my fingers will get tangled as I try to play them. I wonder how you know how long to hold each note." Just then the door opened. Her first lesson was about to begin.

Mrs. Endo introduced herself and then led Sachiko to her studio. "Before we play even one note," she told her new student, "I'm going to teach you about counting. That's something you absolutely must know how to do."

"I could count when I was two," thought Sachiko. "What does counting have to do with music?"

"To read and play music, you need to understand the value of the different kinds of notes," continued Mrs. Endo, as if she could read Sachiko's mind. "The most common types are whole, half, quarter, eighth, and sixteenth notes. A whole note is played and held the longest. It gets the most beats. A half note gets half as many beats as a whole note. A quarter note gets one quarter as many beats as a whole note. You get the idea."

Mrs. Endo pointed out the notes on a chart.

whole note	half note	quarter note	eighth note	sixteenth note
𝅝	𝅗𝅥	♩	♪	𝅘𝅥𝅯

Read•Think•Write If a whole note gets 4 beats, how many beats does a quarter note get?

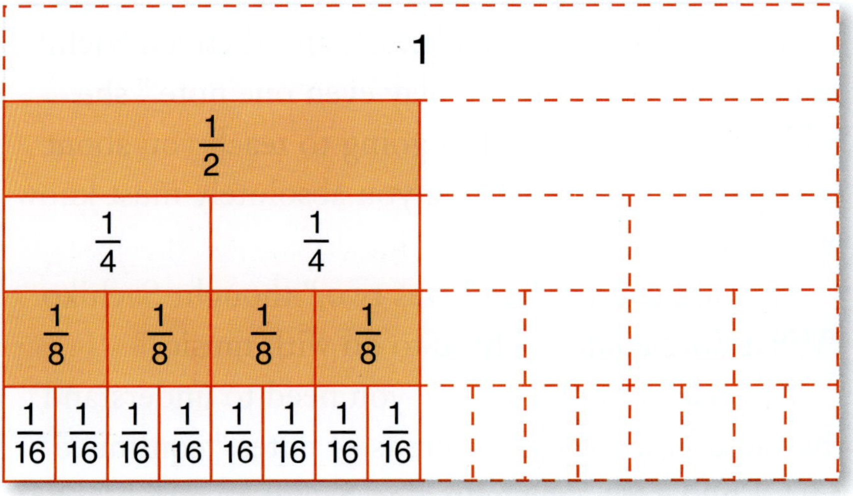

1

$\frac{1}{2}$

$\frac{1}{4}$ $\frac{1}{4}$

$\frac{1}{8}$ $\frac{1}{8}$ $\frac{1}{8}$ $\frac{1}{8}$

$\frac{1}{16}$ $\frac{1}{16}$ $\frac{1}{16}$ $\frac{1}{16}$ $\frac{1}{16}$ $\frac{1}{16}$ $\frac{1}{16}$ $\frac{1}{16}$

"Here's another way to look at it," said Mrs. Endo as she drew a narrow strip on a sheet of paper. "Let's say this represents a whole note. How would we show two half notes?"

"We would divide the strip into two equal parts," Sachiko replied. "Each would be one half of the whole."

"How would we show quarter notes?"

"Divide the strip into fourths," answered Sachiko. They continued with eighth and sixteenth notes.

"I get it!" exclaimed Sachiko. "Your drawings show equivalent fractions. The quarter notes are $\frac{1}{4}$ fraction pieces and the eighth notes are $\frac{1}{8}$ fraction pieces."

Read•Think•Write What fractions are equivalent to $\frac{1}{2}$ and $\frac{2}{4}$? Use eighths and sixteenths in your answer.

“Exactly!” Mrs. Endo responded. “We could write the two quarter notes as the fraction $\frac{2}{4}$. You know that $\frac{2}{4}$ equals $\frac{1}{2}$, so two quarter notes equal one half note. So, if you understand which fractions have the same value, you’ll be able to figure out the rhythm of a piece of music.”

“But how do I know how to count the beats?”

Mrs. Endo drew the fraction strips again. This time, instead of fractions, she drew musical notes. “If the whole note gets held for four beats, each half note gets held for two beats and each quarter note gets held for one beat.”

Mrs. Endo had Sachiko study the strips again. Then she counted out loud and told Sachiko to clap the rhythm for the notes on each strip. Mrs. Endo reminded Sachiko that each strip got four beats altogether. She pointed out that if a strip had more than four notes, Sachiko would need to clap faster to fit in all the notes.

When Sachiko got to the eighth notes, she clapped twice for each beat because two eighth notes equal one quarter note. She continued on to the sixteenth notes.

The clapping pattern was like this. An x stands for each clap.

Beat	1	2	3	4
𝅝 Whole note	x			
𝅗𝅥 Half notes	x		x	
♩ Quarter notes	x	x	x	x
♪ Eighth notes	xx	xx	xx	xx
𝅘𝅥𝅯 Sixteenth notes	xxxx	xxxx	xxxx	xxxx

Read•Think•Write How many times did Sachiko clap for each beat when she got to the sixteenth notes?

“Now it’s time for your homework,” said Mrs. Endo when they had finished. “I want you to take this sheet music home and practice clapping, tapping, or snapping the rhythms as you count out loud. Remember that you hold each whole note for four beats. It will also help if you remember the equivalent fractions we talked about.”

“I didn’t know that counting to four could be so tricky,” laughed Sachiko, “or that music and math were so closely related.”

Read•Think•Write If you play a quarter note for one beat, how long do you play an eighth note?

Responding

Vocabulary

1. How would you show an eighth note as a fraction?
2. Note Important Details What is the denominator of $\frac{1}{8}$?
3. What is the equivalent of one eighth note?
   - A. two quarter notes
   - B. two sixteenth notes
   - C. two half notes
4. Would you hold a thirty-second note a longer time or a shorter time than a sixteenth note?
5. How many thirty-second notes would you play in the same amount of time as one eighth note?

Activity

Work with a partner. Make fraction strips for the following: 1 whole note; 2 half notes; 4 quarter notes; 8 eighth notes. Write the fraction and note on each fraction piece and cut apart the pieces. Show various ways to illustrate four beats.

PHOTOGRAPHY CREDITS: **Cover** © Sudhir Shivaram Nature & Wildlife Photography/Moment/Getty Images; **1** © Canon Boy/Shutterstock; **4** © ian cruickshank/Alamy; **5** © Tierfotoagentur/D.M. Sheldon/Alamy; **7** © Natalia Pryanishnikova/Alamy; **8** ©Arco Images GmbH/Alamy; **10** © Ainars Aunins/Alamy Images; **11** © Peter Hansen/ Shutterstock

Printed in China

ISBN 978-1-328-77206-0

3 4 5 6 7 8 9 10 0940 25 24 23 22 21 20 19 18

4500700961 A B C D E F G

Getting ZZZZs

A head bobs and eyes close. Someone is falling asleep. When it is you, about 10 hours of sleep will give your body a chance to rest. Ten out of 24 hours in a day can be called a $\frac{10}{24}$ day. $\frac{10}{24}$ is a fraction. A fraction is a part of a whole.

Are you snoozing half the day away when you sleep for 10 hours? There are different ways to compare $\frac{10}{24}$ and $\frac{1}{2}$. You can use a bar graph with 24 parts.

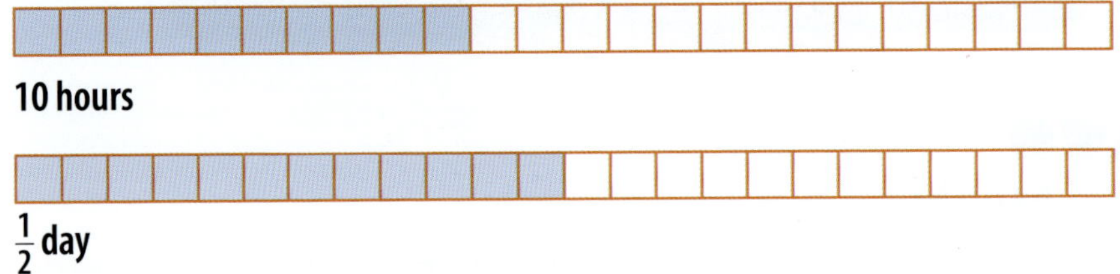

Here is another way. First, rewrite $\frac{1}{2}$ as a fraction with the same denominator as $\frac{10}{24}$. Think: What number do you need to multiply by 2? Because $2 \times 12 = 24$, multiply the denominator 2 by 12. You must then multiply the numerator by the same number, 2. Then, when the denominators are the same, you can compare the numerators.

Figure: $\frac{1}{2} \times \frac{12}{12} = \frac{12}{24}$ **so** $\frac{1}{2} = \frac{12}{24}$ **Compare:** $10 < 12$ **so** $\frac{10}{24} < \frac{12}{24}$

Read·Think·Write How does sleeping 10 hours compare with sleeping half the day?

In this book, you will read about animals and their sleep times, which were measured during research studies. Some animals sleep half the day away. Others sleep fractions of the day such as $\frac{1}{12}$, $\frac{5}{6}$, or $\frac{5}{12}$. Sleeping $\frac{5}{12}$ day is the same as sleeping $\frac{10}{24}$ day, as you do.

You can order $\frac{1}{12}$, $\frac{5}{6}$, and $\frac{5}{12}$ to compare these parts of the day. To order these fractions, first find equivalent fractions with the same denominator. Then you can order the fractions from least to greatest by ordering the numerators. You can also use a number line to order the fractions.

$\frac{5}{6} \times \frac{2}{2} = \frac{10}{12}$ so $\frac{5}{6} = \frac{10}{12}$

$1 < 5 < 10$ so $\frac{1}{12} < \frac{5}{12} < \frac{5}{6}$

0 1

A B C

Read·Think·Write Where would you place $\frac{1}{12}$, $\frac{5}{6}$, and $\frac{5}{12}$ on this number line?

Short Snoozes

Imagine sleeping only about 2, 3, or 4 hours instead of 10. You would probably feel terrible the next morning, and you would need to get more sleep soon! However, sleeping those few hours every day is fine for some animals.

In general, large plant-eating animals, which must be ready to run away from hunting animals, do not sleep much. Horses can lock their legs to keep standing while they nap. Cattle can sleep with their eyes open. Elephants can sleep lightly while they lean against a tree or another elephant. A little noise will quickly awaken them. Elephants in safe places lie down to sleep well. They often snore very loudly!

Read·Think·Write
If a horse sleeps about 3 hours, what fraction of a day does a horse sleep?

Giraffes sleep least, only about 2 hours, or $\frac{2}{24}$ day. Can you think of another fraction that equals $\frac{2}{24}$?

$$\frac{2}{24} \div \frac{2}{2} = \frac{1}{12} \quad \frac{2}{24} = \frac{1}{12}$$

Giraffes sometimes lie down to snooze for a few minutes. A giraffe will rest its head and long neck on its body. Getting up from that position takes time—too long if danger is near. That may be why giraffes often sleep standing up, with their ears twitching and eyes halfway open. Even when giraffes are standing, they usually sleep only a few minutes at a time. The giraffes' total sleep time counts several very short naps.

Half-Day Snoozes

Many animals sleep about half the day. These animals are different sizes and live in different places. Cats, dolphins, and hamsters are just a few of these animals.

Animals Studied	Average Fraction of Day in Sleep
Cats	$\frac{1}{2}$ or $\frac{12}{24}$
Dolphins	$\frac{5}{12}$ or $\frac{10}{24}$
Hamsters	$\frac{7}{12}$ or $\frac{14}{24}$

The average time wild cats or pet cats sleep is about $\frac{1}{2}$ day. The fraction $\frac{1}{2}$ is equivalent to the fraction $\frac{12}{24}$. Cats sleep about 12 hours a day. They do not sleep 12 hours straight. Instead, they take many naps that add up to 12 hours.

Read·Think·Write How many hours are $\frac{5}{12}$ day?

For about $\frac{5}{12}$ day, dolphins sleep as they swim. They can shut down one side of their brain so it is asleep while the other side stays awake! This means that resting dolphins are able to keep swimming and see with one open eye. They can still go to the water's surface and breathe air, so they do not drown.

To find how many hours dolphins snooze on the move, multiply $\frac{5}{12}$ by $\frac{2}{2}$ to find an equivalent fraction with a denominator of 24.

$$\frac{5}{12} \times \frac{2}{2} = \frac{10}{24} \text{ so } \quad \frac{5}{12} = \frac{10}{24}$$

So, dolphins sleep 10 hours a day.

Wild hamsters and pet hamsters make a nest in which to sleep. These small animals are nocturnal, which means that they usually sleep during the day and are awake at night. Since hamsters tend to wake up in the evening, that may be the best time to play with a pet hamster. It is important to remember that hamsters wake grumpy and tend to bite if they are disturbed.

Read·Think·Write Hamsters sleep about $\frac{7}{12}$ day. How many hours are $\frac{7}{12}$ day?

Long Snoozes

Some animals could be called sleeping champs. They usually spend much more time sleeping than doing anything else. While you are reading this book, many long-sleeping animals are snoozing. They go to sleep around sunrise and wake up at night. Some nap during the night, too.

Animals Studied	Average Fraction of Day in Sleep
Owl monkeys	$\frac{2}{3}$ or $\frac{16}{24}$
Armadillos and pythons	$\frac{3}{4}$ or $\frac{18}{24}$
Brown bats	$\frac{5}{6}$ or $\frac{20}{24}$

Owl monkeys are the only monkeys that sleep through the day. They choose "sleeping trees" with lots of thick leaves or vines to snooze in. When owl monkeys sleep $\frac{2}{3}$ day every day, they are sleeping $\frac{2}{3}$ of every week, month, and year, too. The fraction helps to show that owl monkeys spend most of their lives sleeping.

Read·Think·Write How many hours are $\frac{2}{3}$ day? How many hours are $\frac{2}{3}$ week?

Another long-sleeping animal that sleeps in a tree is the python. Each snake curls tightly on a branch, but hangs over it. Armadillos sleep the same amount of time as a python, but they sleep in underground dens.

Do you think you could sleep in a tree? Do you think you could sleep hanging upside down? That's how brown bats sleep. Their toes stay hooked around a bit of a rock, tree, or building.

Which of these long-sleeping animals snoozes the longest? Finding equivalent fractions with the denominator 24 shows the answer!

Owl monkeys	Armadillos and pythons	Brown bats
$\frac{2}{3} \times \frac{8}{8} = \frac{16}{24}$	$\frac{3}{4} \times \frac{6}{6} = \frac{18}{24}$	$\frac{5}{6} \times \frac{4}{4} = \frac{20}{24}$

As you can see, animals sleep in many different positions. Some sleep standing up, and some sleep while swimming. Some sleep in trees, and some sleep underground.

Animals also sleep for different lengths of time. While some sleep for only a few hours a day, others sleep most of the day.

Read·Think·Write Which animal in the book sleeps the same amount of time a person of your age should sleep?

Responding

Math Concepts

1. If an animal sleeps 14 hours, what fraction of the day is that?
2. Order these fractions from least to greatest. (Hint: First find the equivalent fractions and then compare the numerators.)

   $\frac{5}{12}$ $\frac{1}{4}$ $\frac{2}{3}$

3. Which three of these fractions are equivalent?

   $\frac{4}{6}$ $\frac{3}{4}$ $\frac{16}{24}$ $\frac{2}{3}$

Activity

Compare and Contrast The chart lists the times some animals sleep. Write each of these times as a fraction of the day. Then research a favorite animal of yours to find out how much of the day it sleeps.

Name of Animal	Average Amount of Sleep per Day
Elephant	4 hours
Gerbil	13 hours
Squirrel	15 hours
Koala	20 hours